STALIN'S GIANTS
KV-I & KV-II

Horst Scheibert

SCHIFFER MILITARY HISTORY

West Chester, PA

Sources:
Bundesarchiv, Koblenz
Scheibert Archives
Nowarra Archives
Tank School, Bovington Camp, Dorset
Walter Köpfer

Translated from the German by Edward Force.

Printed in the United States of America.
ISBN: 0-88740-404-9

This title was originally published under the title,
Stalins Giganten,
by Podzun-Pallas Verlag, Friedberg.

We are interested in hearing from authors with book ideas on related topics. We are also looking for good photographs in the military history area. We will copy your photographs and credit you should your materials be used in a future Schiffer project.

Published by Schiffer Publishing, Ltd.
1469 Morstein Road
West Chester, Pennsylvania 19380
Please write for a free catalog.
This book may be purchased from the publisher.
Please include $2.95 postage.
Try your bookstore first.

At the northwest exit from Pleskau a KV-1B surprised an antitank company of the 1st Panzer Division (note division emblem on driver's door). What may have happened here on June 5, 1941?

Development of the Klement Woroshilov

Much has been written of the Russian T-34 tank as one of the great surprises of World War II. There are reasons for that, yet another good Russian tank — which was named for a Peoples' Commissar for Defense at the time of the Russian Revolution, Clement Voroshilov (Klimenti Woroshilow, KW or KV for short) — has almost slipped into obscurity in its shadow.

While the T-34 was a medium tank by the standards of the times, the KV ranked among the heavy tanks and became the forerunner of the menacing Joseph Stalin tank, which appeared in the last years of the war and was the direct ancestor of today's T-72 Russian tanks.

As of 1939 Russia had only the T-35 heavy tank, plus a few of its forerunners, the T-32. It was a giant with five turrets and a ten-man (!) crew. It still got into action against the German Wehrmacht in 1941 but had very little effect, since it showed too many weaknesses: Its armor protected only against small-caliber bullets and shrapnel, its steering had technical defects, and its length made it immobile.

And yet possessing a tank that was impervious to large-caliber shells and had more mobility led to two similar developments, the T-100 and the Sergius Mironovich Kirov (SMK). They both had, for the first time, the typical running gear of all later Russian tanks, with torsion-bar suspension, steel road wheels with rubber pads around their hubs, and very wide tracks (68.7mm, as opposed to 30.4 on the T-35). These and the individually sprung wheels allowed a high off-road speed and no longer required aprons to protect the tracks. The reduction to two turrets likewise saved weight, to the benefit of the armor plate.

Above: A T-35 in Lithuania, 1941. It had five turrets (a 7.62 cm L/16, two 4.50 cm, five machine guns) and a ten-man crew. Below: A T-100 with two turrets (one 7.62 cm L/24, one 4.5 cm and three machine guns).

But when these tanks did not prove themselves well in the Finnish War because of their height (3.26 meters) and the laborious operation of their piggyback turrets (requiring a crew of 6 or 7 men), they were developed further into the KV in 1940. It had only one turret, and sparing the weight of the second again allowed heavier armor. It also needed only a five-man crew. It was fitted with a completely new Diesel engine to increase its range (225 instead of 150 km). With its 550 HP (instead of 100 for the T-100), it gave the tank a favorable power-to-weight ratio and thus made it livelier.

On this chassis there was placed either a small turret with a 7.62 cm multipurpose gun or a larger one with a 15.2 cm howitzer, since it was still believed that a support tank (KV-II) was needed along with a pure battle tank (KV-I).

When the war against Russia began, there were barely 600 of the two types together. The German soldiers gave it the nickname "Fat Bello" and feared it particularly because of its invulnerability. Tanks of the KV series took part in all the main battles up to 1943: Leningrad, Moscow, Stalingrad, the Kursk Basin. Originally built only at the Kirov Works in Moscow, after Russian heavy industry was relocated its mass production took place at Panzergrad, near Cheylabinsk in the Urals. By the end of 1943, more than 10,000 of this series had been built.

A KV with a 15.2 cm howitzer, planned as a support tank. It was designated KV-II and is described in this volume, beginning on page 29.

KV-I

The first of them were used successfully in the Finnish winter war of 1939-1940. But only in 1940 were they built and put into service in large numbers. The armor measured 75mm in front (the T-35 had only 30mm there), the hull was welded, and the gun was a 7.62 cm M-1038/1939 (L/30.5) — the same one that was used in the first T-34's. The question immediately arises: Why such a heavy (46.3 tons) tank for "only" a 7.62 cm gun, when the T-34 (only 26.3 tons) carried the same armor, weapons and motor into action with just a four-man crew? So there soon followed a longer 7.62 cm multipurpose gun, and finally an 8.5 cm gun. With this armament the tank was designated KV-85. But when, in 1943, the T-34 was likewise equipped with this 8.5 cm gun (T-34/85), that was the end for this tank, and it was followed as heavy tanks by the Josef Stalin I (JS-I-III) developments with a 12.2 cm L/43 multipurpose gun.

Along with its primary armament, the KV-I had three machine guns (7.62mm): one coaxial type to the right of the gun, a second on the left side of the bow, and another in the rear of the turret.

It cannot be denied that it had its weaknesses:
— The tracks had only a short lifetime and cannot be compared with today's.
— The interior space — like that of the T-34 — was uncomfortable for the crew.
— The commander had only slight visibility, and,
— The training of the crews was — at least in 1941 — far behind that of German armored troops.

At the beginning of the war, and more so during the counter-offensive in the winter of 1941-42, the KV-I and KV-II were driven through the great Russian "front cities" of Moscow and Leningrad.

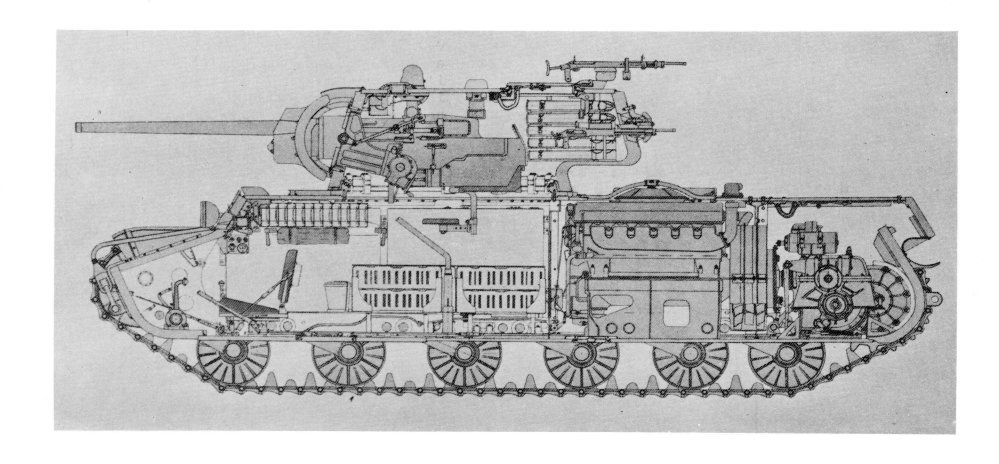

A cross-section drawing of the KV-IC, the KV-IA and IB were similar. The machine gun on the turret roof was added only in 1943. It also had cartridges — not visible here — on the sides of the hull. This meant that direct hits often caused explosions in the fighting compartment, which at least separated the turret from the hull.

Yet thanks to its strong armor and long gun with a velocity of 662.02, it was far superior to the German Panzer IV of the time (1941-42) with its short 75mm L/24 gun.

There were several other versions, though, before the KV-85:

KV-IA

It differed from the first series in having a longer (L/41.5) gun, stronger charges in the shells, and road wheels in somewhat changed form. It was built in 1940, after the first experiences in the Finnish war, and earlier models were equipped with the new gun.

The photo at lower right clearly shows the longer gun of the KV-IA. Below, an explosion has torn the KV-I apart. The turret and parts of the engine cover lie beside the hull. The KV-I at upper right was hit on one of the roads leading out of Leningrad.

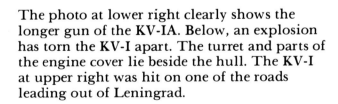

This KV-IA was probably hit on the right side, where the bowed-out track apron was attached. The hole by the covered gun cradle is for the coaxial machine gun. The two horns on the forward part of the turret cover are panoramic scopes. This tank was destroyed on the third day of combat on the Dubyssa in Lithuania (near Rossienie).

This tank was also torn apart by an explosion of its own ammunition after being hit by an 88mm Flak shell near Gauri, Estonia. The tank is seen here from the rear. It has an auxiliary fuel tank on the track apron. The machine gun in a ball mantelet at the rear of the turret can be seen.

Above: A fully destroyed KV-I. Below: Another view of the tank shown on the inside front cover. The wide tracks, with good traction, are very clearly seen.

Above: Two destroyed KV-I tanks. Below: The shell size (one lies in the foreground) and the rear machine gun can be seen. These heavy tanks were always interesting things to examine.

Above: A KV-IA had just rolled over a 50mm antitank gun when it met its fate (Demyansk, 1941-42). Below: this KV-IA with auxiliary tanks got stuck. Like the tank above, its turret is in the 6:00 position.

Many Russian tanks, including the KV, were captured while still on railroad cars, not yet unloaded. This picture was taken during the battle for Smolensk.

This KV-IA shows the shape of its turret very clearly. The lettering on it says "Shapayev", the name of a Russian cavalry general in the Revolution. Many KV's were painted with the names of Soviet heroes; later names from Russia's earlier history were also used.

In the early pocket battles, many KV's were abandoned when they ran out of fuel. Despite preparation, they often could not be towed away, so most of them fell into German hands unharmed. Note the lack of a bow machine gun on this one.

KV-IB

It was a **KV-IA** with additional armor plates, 25 to 35mm thick, screwed onto the turret and welded to the bow and driver's area. It is easy to recognize by the many screw-heads on the turret.

As of 1942 the KV-I was fitted with a cast armored turret. This KV tank was designated **KV-IB**. The turret is easily told apart from the welded one of the first version, and the rear of the turret does not project so much.

KV-IC

In 1943 a new cast turret was made, now with 120mm armor. The hull was also 90mm thick. Its motor was strengthened to 600 HP, and the tracks widened to 70 cm. Most of the KV's made were of this type. It can only be told from the **KV-IB** (cast turret) with difficulty.

Above: The KV-IB is easy to recognize by its screwed-on additional armor plates. A few KV-IA's with the shorter gun barrel were also fitted with these extra plates. On the picture at right one can see the mark made by a non-penetrating hit (probably from a 37mm antitank gun). The cutouts on the turret sides allowed the original peepholes and pistol loopholes to be used again. The latter could be closed from inside by a plug on a chain.

14

A KV-IB stopped by track damage (direct hit?). Thanks to the screwed-on or welded armor plates, the tank had 110mm (!) armor on the front of the hull and turret. But its weight increased by four tons to 47.5 tons, and thus its power-to-weight ratio fell from 12.6 to 11.6 HP per ton.

One of the many KV's that broke down or were destroyed in the Baltics or Ingermanland — here before Leningrad in 1941. The Kinon blocks seen on the turret allowed a view to the sides. They were also on the rear of the turret for the (semi-prone!) rear machine gunner and the driver. The latter are recognizable on the next picture.

This **KV-IB** was also prepared to be towed away. But it was not towed; the German advance was faster. The welded-on additional armor plates on the rear of the turret and the driver's area can be seen clearly here. A German soldier has drawn a cross on the rear of the turret with chalk.

The only weapon that was a match for the KV, and even proved to be superior, was the 88mm Flak gun (above) used in ground combat. After the first surprises — individual KV's blocked advance routes for up to thirty hours — they were issued to every battle group. But the KV was able to flatten motorcycles (left), 37mm antitank guns (right) and other equipment here and there.

One of the two **KV-IB's** near Gauri (Estonia) that shocked the spearheads of the 6th Panzer Division on the Dünaburg-Pleskau road and could only be put out of action when 88mm Flak guns were brought into action. The second KV is still burning in the background.

One of the two KV-IB's from Gauri (Estonia) shows where the 88mm shell penetrated (upper left). The mouth of the machine gun can be recognized to the right of the armored gun cradle.

This close-up shot shows the peephole and pistol loophole of a KV-IB. It is also the place where the two additional turret armor plates meet.

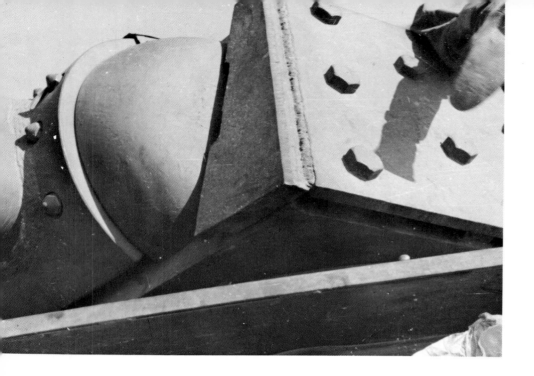

Upper left: It can be seen that the almost triangular welded-on left armor plate at the front of the turret is attached to the screwed-on side armor by a welded seam.

Above: Since the right track apron is missing, it can be seen here how far the part of the KV-IB's fighting compartment inside the hull has been strengthened by additional welded-on plates.

Left: A shot-down KV-IB shows where it was hit. The middle return roller has been blown away.

A KV-IB with cast turret. It can scarcely — often not at all — be distinguished from the KV-IC in photos. The latter had a more rounded turret rear, heavier armor and a somewhat more powerful motor. Compared to the KV-IC, rarely few KV-IB's (with cast turrets) were made.

Right page: Here are four pictures showing the KV-IC from all sides. The four Kinon blocks on the turret roof, the rack for an anti-aircraft machine gun, the ventilator for the fighting compartment (between the two panoramic scopes), the changed engine cover and the additional armoring that sticks up over the driver's area are easy to see. But these photos also show a weak point in the tank: for the five crewmen there were only two hatches. That meant slow getting in and out. In crises the results were often fatal.

Above: At the Don bend in 1942. The inscription on the left tank (KV-IB or KV-IC) says "Kutuzov's Heir." Michael Kutuzov (1745-1813) was a Russian field marshal who led a Russian-Austrian army at the battle of Austerlitz (1805) and came out of the battle of Smolensk (1812) victorious. It is interesting that, after Stalin proclaimed the "Great Homeland War" in 1942, the Russian tanks also bore the names of Czarist heroes.

Below: A KV-IC in the spring mud near Rshev, 1942.

On all three of these camouflage-painted KV-I tanks is the lettering: Moscow Kolchose Workers. They were so named because the Kolchose workers from the Moscow area had contributed money with which they could be built. Here Kolchose workers greet the crews of the tanks. Note the new commanders' cupolas with panoramic views on these tanks. Yet the left panoramic scope for the gunner is still present on the KV-I. It was also retained on the KV-85. Each of these tanks also bears the four auxiliary fuel tanks that were obligatory as of 1943 on the rear track aprons. They increased the range from 250 kilometers (on the road) to 400.

The helmets of the Russian tank gunners have three rubber pads on top and headphones in the earflaps.

KV-85

It was the urgent further development of the KV-I and differed from it in having a larger turret with a better shell-deflecting shape, the already noted 8.5 cm M-1943 D-5 T 85 L/51.5 multipurpose gun (also used in the T-34/85 and SU-85), and a new commander's cupola. It reached the front in the summer of 1943 and turned up in large numbers at the battle around Kursk (Citadel). It carried a crew of only four soldiers. Available KV-IC's were upgraded to equal it (winter 1943-44).

At the start of 1944 production of all KV models was halted and only the "Josef Stalin" built; its chassis differed only slightly from that of the KV.

KV-8

A variant of the KV-IC. Instead of a gun, it had a 4.5 cm flamethrower.

KV-Is

A KV-I made in very small numbers, built for speed (the "s" stands for skorostnoy = speed) by reducing its weight. Here for the first time, as later in the kW-85, the commander was given his own observation cupola.

Both photos on this page show the KV-85. It saw service particularly in 1943, but was replaced more and more by the Josef Stalin series in 1944.

Above: **A KV-IIA.** As powerful as it appeared, its design was a failure, neither a tank nor an armored howitzer. For the latter use it did not need a rotating turret, but did need appropriate devices for indirect aiming, which it lacked.

Right page: A captured KV-IIA rolls through a German city.

KV-II

The KV-II appeared in 1940, and was described as a support or artillery tank — it had a six man crew. Since its boxy turret could rotate 360 degrees and thus could be fired when aimed directly, it is regarded as member of the tank family and not as armored artillery — in spite of its howitzer. It did not prove itself in combat, since its heavy, boxy turret could only be turned on completely flat ground. Its main armament was a 15.2 cm M-1938/39 L/20 howitzer. Its shells were like those of the towed howitzer of the same caliber, but were used here in bullet form. In addition, it had two machine guns, one in the bow, the other in the rear of the turret. With its heavy turret, its weight amounted to 57 tons, and its power-to-weight ratio sank to 9.45 HP per ton (in comparison, the KV-I had 12.6 HP, and the present-day Bundeswehr Leopard has a 20.8 HP per ton). It was said by the Russians that it proved itself well when attacking bunkers on the Mannerheim Line during the Finnish winter war of 1940. In battle against the Wehrmacht though, it was effective only on account of its size, heavy armament, and the ability to use its weight and wide tracks to flatten motorcycles and trucks. However, when confronted with a German 88mm gun, its fate was sealed quickly because of its slow speed. All in all, it proved to be a faulty design, and its construction was therefore halted at the end of 1941.

This is how it looked when it attacked, but its big silhouette gave the 88mm Flak guns an easy target to hit.

And for technical reasons, and organizational as well, for sometimes only fuel was lacking, many of them were abandoned.

A **KV-IIA** with a bullet hole in its turret. One of its shells can be seen on top. The opened hatches were part of the engine cover and were installed for servicing.

It was more impressive because of its large size, loud noise and invulnerability than for any real success. It was too slow, too inaccurate, and needed flat ground to be able to turn the turret properly.

A KV-IA with a bullet hole in its turret. German tank gunners are having a look at the other side's technology.

Above and below are the KV-IIA tanks with combat tracks. The smaller holes were made by 37 and 50mm antitank guns. The rear entrance hatch of the turret is easy to see in the upper picture.

This tank also shows one of its 15.2 cm shells on its turret roof. The road wheels of the KV tanks did not have their rubber pads outside on the rims (as did the German tanks), but around the center of the wheel. The wheels were made of steel — which made for loud road noise.

An 88mm direct hit on its right side was its undoing. It also tore up the track and the apron. It shows a different type of shell (on the rear apron) from the tank on the previous page. A 15.2 cm shell casing stands on the ground at the right front.

A KV-IIA in a swamp at the edge of an airfield. A Junkers Ju 52 can be recognized on the horizon and, between it and the tank, a Russian plane that belongs in a museum.

Left: If a direct hit set off an ammunition explosion, the tank usually burst apart, and the turret at least came off the hull.

Right page: A KV-IIA with battle scars. One disadvantage of this heavy Russian tank was that its crews were trained too hastily. They often got stuck, which made the antitank guns' job easier.

Here is a fully exploded KV-IIA on the Stalin Line at the Estonian-Russian border. There the Stalin Line consisted of bunkers and battle stations, some of which were still under construction. Since the KV was built in Leningrad at the beginning of the war, it was widely used in the Baltic area in 1941 and later in the defense of Leningrad, often crewed by workers from the tank factory.

Left page: A KV-IIA on display during the war.

Right: A battle-scarred tank. The arrangement of the rear turret door can be seen here — it had no machine gun. On the other hand, the KV-II in the picture below shows a rear machine gun.

In the picture at lower right, a shell and casing can be seen.

All the photos on these two pages show the KV-IIA. At lower right, a tank captured near Volkov still bears its winter paint, plus a German cross on the turret. Shot-up or abandoned enemy tanks were often used by German advanced artillery observers as observation posts. The KV-II weighed almost 60 tons (fighting weight) and so could not use most bridges, and they often got stuck while fording the many rivers. Its cruising speed was 22 kph; off the road it could do no better than 12 kph. Thus it was only half as fast as the German tanks — a slow-moving monster! With the low muzzle velocity of its shells, it was useless for fighting enemy tanks. All that resulted in the cancellation of its pointless development, and so the KV-IIA — like the even older T-35 — was rarely seen after the summer offensive of 1941.

Right page: A KV-IIA with white winter paint. It got stuck in a swamp. The bundle of logs did not help it get out; it fell into German hands.

In comparison to the KV-I, it had one more crewman. Its heaviest armor was 100mm thick, but since it was incapable of fighting against tanks, this heavy armor was unnecessary. As an armored self-propelled howitzer firing out of hidden positions, shrapnel protection would have sufficed.

Tools of all kinds were carried in the flat sheet-metal boxes on the aprons. They are usually broken because anybody could use their contents.

The following versions also existed:

KV-IIA

As described above.

KV-IIB

It had the somewhat improved chassis of the KV-IB and a slightly modified turret. The latter could be recognized by a different rounded shield and a pointed turret rear. There were only small numbers of it built, and even fewer of a flamethrower version.

KV-II.2

In 1943 a 12.2 cm antitank gun was mounted on a KV chassis in a rotating turret and counted — probably because of the large caliber — as a member of the KV-II series. But only a few test vehicles were built.

The KV-IIB is shown on these two pages. It had a different rounded shield, a two-section rear turret, was 50 cm taller (4.17 meters!), and no machine gun in the driver's area. All in all, it was no improvement — quite the opposite.

A good rear view of a KV-IIB with the engine-room servicing hatches and the Kinon blocks above the pointed turret rear.

Technical Data

KV-I C

Crew	5
Fighting weight	47.0 tons
Overall length	6.80 meters
Length minus gun	6.75 meters
Overall width	3.33 meters
Width over tracks	3.24 meters
Width mid-track	2.64 meters
Overall height	3.25 meters
Ground clearance	37 cm
Track extent	4.20 meters
Top speed on road	29 kph
Off-road speed	12 kph
Cruising speed	24 kph
Fuel capacity	544 liters
Range on road	250 km
Range off road	176 km
Turning circle	9.45 meters
Power-to-weight ratio	11.7 HP/ton
Ground pressure	0.79 kg/sq. cm
Ditch spanning	2.80 meters
Vertical step	0.91 meters
Climbing ability	36 degrees
Fording ability (unassisted)	1.44 meters
Motor model	W-2 K
Motor type	V-12 Diesel
Horsepower/rpm	550/2150
Cooling	Water
Transmission	Variable gears
Speeds forward/reverse	5/1
Steering	Clutch & brake
Track type	Cast manganese steel, central horn on every other plate
Track width	68.7 cm
Track gap	15.6 cm
Links per track	87-90
Suspension	1 torsion bar per road wheel at equal distances from leading arm
Wheels per side	6 double road wheels
Primary armament	7.62 cm M-1940 L/41.5 multipurpose gun
Secondary armament	3 7.62mm DT machine guns
Turret rotation arc	360 degrees
Elevation	-4/+24.5 degrees
Primary ammunition	114 rounds
Secondary ammunition	3024 rounds
Communication	Radio & on-board speaker
Armor, turret	40 to 120mm
Hull side, front	90 + 40mm
Hull side, rear	90
Hull, upper front	75 + 35mm
Hull, rear	75mm
Hull, floor	35mm
Hull covering	40

KV-II B

Crew	6
Fighting weight	57 tons
Overall length	6.80 meters
Length minus gun	6.77 meters
Overall width	3.33 meters
Width over tracks	3.24 meters
Width mid-track	2.64 meters
Overall height	4.17 meters
Ground clearance	36.7 cm
Chain extent	4.39 meters
Top speed on road	25 kph
Speed off road	12 kph
Cruising speed	22 kph
Fuel capacity	590 liters
Range on road	160 km
Range off road	134 km
Turning circle	9.45 meters
Power-to-weight ratio	9.7 HP/ton
Ground pressure	0.86 kg/sq. cm
Ditch crossing	2.80 meters
Vertical step	0.91 meters
Climbing ability	34 degrees
Fording ability (unassisted)	1.45 meters
Motor model	W-2 K
Motor type	V-12 Diesel
Horsepower/rpm	550/2150
Cooling	Water
Transmission	Variable gears
Speeds forward/reverse	5/1
Steering	Clutch & brake
Track type	cast manganese steel, central horn on every other plate
Track width	68.7 cm
Track gap	15.6 cm
Links per track	87-90
Suspension	1 torsion bar per road wheel at equal distances from leading arm
Wheels per side	6 double road wheels per side
Primary armament	15.2 cm M01938/40 L/20 (M-10) howitzer
Secondary armament	2 7.62mm DT machine guns
Turret rotation arc	360 degrees
Elevation arc	-4 to +24.5 degrees
Primary ammunition	36 rounds
Secondary ammunition	3087 rounds
Communication	Radio & on-board speaker
Armor, turret	35 to 100mm
Hull side, front	75 + 35
Hull side, rear	75
Hull, upper front	75 + 35
Hull, rear	75
Hull floor	35
Hull cover	35

A KV-IIB in Latvia (1941), destroyed by units of the 1st Panzer Division.

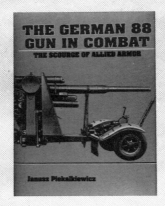
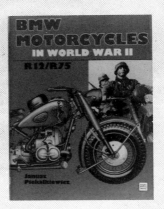